I0011071

TABLE OF CONTENTS

WHAT IS THE AMAZON TAP?

What is the Amazon Tap device? Why should you consider purchasing one? What makes the Amazon Tap different from other Amazon Alexa-enable products? These questions and more will be answered in this chapter.

Basic Amazon Tap Information

Firstly, the Amazon Tap is part of the Alexa family, meaning that it offers some of the same services as the Echo, Echo Dot, and Fire TV. The Tap's main function is to stream music through its excellent sound system. This device provides 360-degree surround sound and is easily portable. It is a party speaker that can provide more sound than you would expect from its small frame.

The Amazon Tap is a bit pricier than the Echo Dot but less expensive than the Echo. It offers a great, middle-of-the-road option when it comes to price. However, the Tap also offers a lot of great advantages that other Bluetooth speakers do not. Here are a few of the options that the Tap has enabled:

- The Amazon Tap works with mobile hotspots as well as wi-fi while the other Bluetooth speakers mainly work with wi-fi and wi-fi only.

- Obviously, the Amazon Tap offers voice control options with Alexa that other Bluetooth speakers do not provide. While Amazon is allowing more and more devices to imbed Alexa in their devices so that the devices can communicate with each other, other Bluetooth speakers have not received that feature yet. You will be able to ask questions, change the song, or stop the music streaming with your voice.

- Last of all, the Amazon Tap offers Dolby while many other Bluetooth speakers do not. For those of you who do not know electronics intimately, you may be asking what Dolby is. Dolby is a noise-reduction system that is used in listening to your voice and in playing music to keep static and hissing down to a minimum. Offering Dolby means that the Amazon Tap is giving you high-quality playback.

Here are a few technical points that distinguish the Amazon Tap. The Amazon Tap weighs 16.6 oz or 470 grams. In comparison, the Amazon Echo weighs in at 37.5 ounces and the Amazon Echo Dot weighs in at 5.7 ounces, a little less than half the weight of the Amazon Tap. A charging cradle is included in the Amazon Tap package. The battery included in the Tap has the potential to last up to nine hours depending on your usage. However, you can also choose to plug the

Tap into an outlet and use it as it is charging. The Taps dependence on battery, not electricity, means that it is truly a portable speaker.

The Tap contains dual 1.5 inch drivers and dual passive radiators in case you want to extend the bass's sound. Additionally, the Tap comes with a one year warranty, and you will be given the opportunity to extend the warranty, should you so choose.

What Makes the Amazon Tap Stand Out?

If you have read the other manuals written on this subject, you know that the Amazon Echo and Amazon Echo Dot are both devices that stay in one place while you are using them. You can take them to a new place and use them there; however, they need to be connected to Wi-fi in order to access Alexa and other features. You will need to find a nearby plug; the Echo and Echo Dot are not devices that can be used on-the-go.

The Tap is powered by its battery, meaning that if it is charged, you don't need to have it plugged in. Because the Tap does not need to be plugged in and you can connect it to mobile hotspots, you can virtually transport this device to any place you can get cellphone signal. You can turn on your cellphone's mobile hotspot feature and connect your Tap to the hotspot. This means that you can stream music on an excellent speaker when you are outside- at the park, at the beach, at a picnic. Your imagination is your limit.

Additionally, because the Tap is built with Dolby, the music does not get fuzzy when you turn it up. Many speakers can become fuzzy and unclear the louder you turn up the volume. The Tap keeps streaming your music clearly no matter how high you turn it.

If you specifically want to have Alexa in your devices, but you also want to have a great speaker, then you are left with the Amazon Echo and the Tap. The Tap is the cheaper of those options, giving you both a great speaker and the Alexa feature. The Echo Dot gives you the Alexa feature, but it does not have a music built for streaming music.

The last main difference between the Tap and its Echo counterparts that support Alexa is that the Tap must have the button pressed to hear you when you speak. This is advantageous for those of you who do not want the Tap listening 100% of the time. There have been some privacy concerns about having the Echo or Echo Dot listening all the time, so the Tap addresses those concerns. It will only hear you when the button is pressed. This is similar to the Amazon Fire TV remote that has the Alexa voice feature. However, with the Amazon Fire TV, you need your TV to be on in order to hear and respond to your commands. The Tap does not need any other devices on or present in order to respond to your question.

Conclusion

The Tap is an affordable version of the Amazon Echo, running at $89.99. The Tap is activated by a button, not by your voice. The Tap runs on battery life and can be taken to various place to stream music via wi-fi or mobile hotspots. Owning a Tap will give you the fun features of the Alexa voice system, and it will also help you enjoy your music in a variety of places.

If you already have an Amazon Tap and you want to begin using its great features, then go ahead to chapter two to learn about how to set up your device.

SETTING UP THE AMAZON TAP

If this is your first Alexa device, then welcome to the world of voice control! If this is your second or third or fourth Alexa device, then you already know how life-changing these devices can be. In this chapter, you will find information in the following categories: what comes in your box, how to set up your Tap, and how to get to know your device.

What Comes in Your Box

Your box should contain five items. When you receive your Amazon Tap, go ahead and open the box. Make sure that everything you are supposed to receive is in the box and in good working order. The items that you should find in the box are the following:

- Amazon Tap- This is the main device, and you will learn more about its buttons and features later on.
- Charging Cradle- The charging cradle is a circular piece that is used to keep your Amazon Tap in place while you are charging it.
- Power Adaptor and Charging Cable- These are two pieces. One is the wall charger, and the other is the cord that will connect the charging cradle to the wall charger.

You can choose to charge the Tap in your car if you have a car charger, but the car charger does not come with this package.

- Instructions- You should have a quick start guide in your box. This will give you some basic information about your Tap.

You can choose to purchase a cover for your Tap. The Tap covers are called "slings" and can provide protection for your device when you take it to multiple places.

Get To Know Your Tap

Before you begin using your Tap, you should get to know the buttons and their uses. This section will briefly introduce the buttons, their locations, and their uses to you.

Firstly, you have the microphone button which is in the front and center of your Tap. This is the button you will press to speak. Remember that the Tap will not hear your request unless you are pushing the button.

Next, you should see a few lights on the front of your Tap. They won't be doing anything when you open it, but they do have a few different purposes. The lights will be in four different colors: red, cyan, blue, and amber. If you see an amber light that is moving from left to right, this means that your Amazon Tap is setting up and is ready to be connected to the Alexa App. If red lights begin pulsing, that means that the Tap was not able to fulfill your request. This could be due to inability to connect to the internet or inability to hear your voice quickly. If you see blue lights pulsing over unlit lights, that means your Tap is going into pairing mode to be connected to Bluetooth speakers.

If you are pressing the microphone button, you should see cyan lights filling the indicator lights. This means that the Tap is listening and ready to hear your request. Once those same cyan lights flash, it means that the Tap is fulfilling your request. Don't worry about knowing what all of these light signals mean as you will get used to them as you begin playing with the Tap.

On the top of your Tap, you will have several clear buttons that are used to play, pause, go to the previous or next item in the song list, and change the volume. On the back near the bottom, you should spot another small set of buttons. The top one is the power button and should be held down for several seconds to start your Tap. You will then see 3.5mm audio port. You can use this audio port to connect even bigger speakers if you wish to do so. A 3.5 mm audio cable does not come with the Tap device.

Underneath the audio port, you will see the place where your charger will be inserted. Below the micro-USB port, you will find the Wi-fi/ Bluetooth button.

Steps for Setting Up Your Tap

Here are the steps you should follow when setting up your Tap. Remember that your Tap does need time to charge, so you should not plan to use it the moment you open the package.

1. **Charge your Tap.** The Amazon Tap takes about four hours to charge. Because it gives you nine hours of playback time, this is a great charging time. Go ahead and use your charging cable and power adaptor to plug the charging cradle into the wall. Place your Tap into the charging cradle, and start charging the battery. You can also charge the Tap by plugging it into the port on the back of the device. The power button should glow to alert you that you have plugged the device in correctly and that it is charging. Once the device has enough battery, the lights should glow amber, and Alexa will greet you.

2. **Download the Alexa App.** The Alexa App is a free app that is used to help control all Alexa devices. Even though you can control many parts of the device with

your voice only, the app allows you to have a way to look at and change controls immediately. There are also some cases where the Tap literally cannot complete your command, and you will need the Alexa App. You can download this app on a Kindle 2.0 or more, Android 4.4 or higher, and iOS 8.0 or higher. You can find this app for free in the Apple app store, Google play store, or the Amazon app store.

3. **Connect to Wi-fi.** To access Alexa's full capabilities, you will need to connect your device to internet. Here are the steps you can follow to do so. If you are connecting to a wi-fi network, you will go to the settings selection in your Amazon App and select **Update Wi-fi.** Hold down the wi-fi/ Bluetooth button on the back of your Tap until you see a red light glow on the front. This should connect your Tap to the Alexa App, though you may need to enter the **Set up a New Device** section to enable the Alexa App to talk to your Tap. You should see your wi-fi network on the list of networks. Enter any passwords that may be necessary. If your wi-fi is a new network, you may need to rescan the area or add a network. Once you are able to locate your network, you can press the connect button, and you should be able to speak with Alexa normally.

4. **Connect to a Mobile Hotspot.** You could also choose to connect to a mobile hotspot. Doing these steps beforehand will help your Tap remember the phone for future use. Use your mobile device to open the mobile hotspot. There should be a name and password for your mobile hotspot. Go into the settings area of your Alexa App and select **Update Wi-fi.** Hold down for at least five seconds the Wi-fi/ Bluetooth button on the back of your Amazon Tap. You should see a selection that says: **Use this Device as a Wi-fi Hotspot.** You will press **Start** then enter the name and password given you by your phone. You should be able to connect. This connection is controlled by your phone and limited by your phone's capability of being a hotspot.

5. **Talk to Alexa.** Now that you have finished setting up your Tap, you can begin talking to Alexa. Remember to press that microphone button. Unlike the Echo and Echo Dot, you do not need to say the word "Alexa" before each question. You might choose to ask "how much battery is left?" to see how much more time you need to spend charging your Tap before you can use it.

6. **Sleep Mode.** You can choose to put your Tap in sleep mode if you will not be using it for a period of time. Simply press and hold down the power button until the

indicator lights change color. This is a great way to save battery.

Now, you are ready to begin using your Tap to stream music and change your everyday life. In the next chapter, you will find out about voice controls that you can use with your Amazon Tap.

VOICE CONTROL FOR DEVICES

You can connect multiple devices with your Alexa-powered Amazon Tap. There are an array of smart home products that are now being created with Alexa's voice. Before you are able to control your different devices using your voice, you must connect them. Next, you will need to learn the commands that will extract the information or cause the change that you want. This chapter will specifically help you connect different devices you may have or want to have to your Amazon Tap so that you can control them all with just the tap of a button and a short command.

Connecting Your Smart Thermostat

Amazon Tap devices can be used as smart home hubs. However, many times, you may need to have a separate smart home hub in order to connect your smart home devices to your Alexa-powered Tap. This section will help you connect any of the compatible smart thermostats to your Tap. You will also learn basic voice commands that you can give to your Tap that will work with your thermostat.

If your thermostat can be connected directly to your Amazon Tap without needing a smart home hub, you will follow these steps. First, you will make sure that thermostat has been installed properly according to its directions. Once the thermostat is up and running, you can proceed. Open the Alexa App on your smartphone or tablet and select **Smart Home.** It should be about halfway down the list.

A list of options that are available should appear. If you have multiple smart home devices, you might see multiple ones pop up. Select the name of your device and press **Continue.** You will then need to log into the account that you created with your smart home thermostat. A screen should appear saying that you have been successful in linking Alexa with your thermostat. You will then select **Discover Devices.** Your Alexa app should be able to find and locate your thermostat now, and the thermostat should appear in your list of devices.

There are quite a few smart thermostats that can now be controlled by Alexa's voice. Some of those are the Nest, the Sensi, and the Ecobee3. If you are looking for a cheap way to set up your smart home with your Amazon Tap, then you can choose the Sensi, which is priced at $125 right now. However, the Ecobee3 offers quite a few more features such as the ability to use algorithms to take the outside temperature into account when setting your heating and cooling inside the house. However, this smart thermostat is also $249 right now, double the price of the Sensi. If you are worried about price, then you should select the Sensi. If price is not a factor, then you should select the Ecobee3.

Now that the thermostat is linked to your Tap, you can give some of the following commands. Your thermostat should have a name that you selected. It may simply be the brand of the thermostat you have, or it may be the name of the room in which you placed the thermostat. If you have multiple thermostats, then you will need to use the name of the thermostat in place of the word "thermostat" in the commands below.

- "Turn the temperature up." That will adjust the temperature up by two degrees.
- "Turn the temperature down." That will adjust the temperature down by two degrees.
- "Raise/ Lower the temperature by five degrees." You can raise or lower the temperature by a set amount of degrees by substituting in the number you want in place of five.
- "Set the temperature to seventy-six degrees." You can put in any number, and the thermostat will change however much necessary to reach that level. Remember that if you have multiple thermostats, you will need to specify. "Set Downstairs Thermostat to seventy-six degrees, for example."

This is how you will connect and command your smart home thermostat.

Connect to Your Harmony Remote

Another device which you can control using the voice controls in Tap is the Harmony Remote. You will only be able to control activities that you have already created on the Harmony Remote, not control or create any new activities using Alexa. To control the Harmony Remote, you will need the IFTTT app.

Your first step is to connect your IFTTT account to your Amazon account. Make sure that your Harmony Elite account is connected to your IFTTT account. Once all these accounts are connected, you will be able to use voice commands to control your Harmony Remote. That app is what helps Alexa "talk" to the device. Open your IFTTT account and choose the icon that is located in the upper right corner of the screen.

The screen should now say "My Recipes" at the top of your device. Press the plus icon. You should see a button at the bottom of the screen that says "Create a Recipe." Select that button. You will see a sentence appear on your screen that says "If + then +." You can add two items into that equation to create a recipe.

Select the first plus button and scroll through the apps until you reach the Amazon Alexa App. Choose "say a specific phrase." It will give you a list of options, or you can type your own. In the case of the

Harmony Remote, you might choose the phrase "dish tv." This is not a complete sentence, but a phrase that will catch the device's attention.

Go back to your "If + then +" screen and select the second plus mark. Find the Harmony app icon. You can choose whether the phrase you just typed or selected in the previous step will start or end the activity. You will be able to select the Harmony Remote and finish your recipe. Now, you will be able to use the voice controls that you typed into the "if" space to control your Harmony remote. This is a really neat feature as it allows you to customize your commands. You can create multiple other commands as well that allow the app to be more versatile in recognizing your commands.

This device ranges in price, depending on which Harmony Remote you want. Some come in package deals with a hub. Other times, the hub is sold separately. The remote is normally no less than $150 and no more than $250. I would not recommend you buy this remote simply to use it with Alexa. However, if you already have the remote, then the Tap's voice control features make a great addition.

One great example of using IFTTT is to trigger the ring feature on your phone whenever you say "Alexa, find my phone".

Onvocal Bluetooth Earphones

These headphones are a new product and one of the only sets that works with Alexa. They are great because they allow Alexa to answer your questions directly into your ears. It also allows you to have more control. These earphones have a button that when pressed will allow you to give commands to Alexa and receive answers. All of the voice controls that you speak using your Tap can be spoken using the headphones, and their attachment is simple.

Additionally, the earphones are designed to sit around your head comfortably and not pinch or warm your ears. With one press of the button, you can speak your command and have Alexa answer just as she does with your Tap.

The Onvocal Bluetooth Earphones are a fairly new invention and will not be available to ship until January; however, you can preorder them for $399. Continue reading in the following chapter for more information on how to listen to music and media using your Tap.

MUSIC AND MEDIA ON YOUR TAP

Because the Amazon Tap has such a great speaker, you will want to make use of streaming music. Additionally, because the Tap is portable and can use mobile hotspots, you will be able to take it with you wherever you go. In this chapter, you will learn how to stream Amazon Music Unlimited, how to make a third-party music streaming system your default, and how to listen to audiobooks.

Amazon Music Unlimited

You can choose to have an Amazon Music Unlimited subscription if you want access to millions of songs. Amazon Prime does offer free songs to Prime members; however, the selection is limited. If you want even more selection and you are a frequent Amazon user, looking into purchasing an Amazon Music Unlimited account may be the right decision for you. To see if Amazon Music Unlimited is the right decision for you, you will have access to a month's free trial.

You can start the free trial or start your membership today by going to the Amazon website and selecting **Get Amazon Music Unlimited.** If you have never had a free trial before, you will automatically be started with the free trial before you need to begin paying. You will need to enter your basic information and billing numbers. Then, you will be able to confirm and start your subscription. If you have a Prime membership, you will be eligible for a discount price on your Amazon Music Unlimited purchase. The price for Prime members is $7.99 per month, and the price for non-Prime members is $9.99 per month. If you only plan on using your Amazon Music Unlimited streaming capabilities for your Tap, then you can select an even less expensive plan at $3.99 per month. However, if you want to have Amazon Music Unlimited on your phone or tablet as well, then you will need to select the $7.99 plan.

Once you have signed up for your Amazon Music Unlimited account, you will be able to access it from your tap. By using the command "Sign up for Amazon Music Unlimited" to your Amazon Tap, you will use the "1-click" order form and be able to order the subscription with your voice only.

Here are sample commands that you can use to play music:

"Play (song name) by (artist name)."

"Play (album name) by (artist name)."

"Play (playlist name)."

"Play songs by (artist name)."

By using these commands, you can stream music from your Amazon Prime or Amazon Music Unlimited account. If you want to stream music from a third-party provider, if you already have an account with Spotify, for example, then you need to follow the steps in the next section.

Streaming Music Using a Third Party System

If you already have a Spotify of Pandora subscription, or if you simply do not mind the adds, then you may want to make their music your default music. By making one of these your default music provider, you won't need to say "Play (song name) by (artist name) on Spotify." You can simply ask for the song, and it will begin playing from Spotify, if you set Spotify as your default.

First, you need to go to your Alexa App and open the settings menu. Look for **Music and Media.** Select that button then select **Choose Default Music Services.** You should see music library and music streaming options available. You should be able to see your selection available. Click the little bubble next to the name of the service that you want to be your default. Now, you can give the same commands described in the above section, but they will be directed to the service you selected instead.

Listen to Audiobooks

Your Tap is also capable of reading audiobooks to you and not just streaming music. Alexa supports an application called Whispersync for Voice. One of the amazing things about using audiobooks with Alexa is that you can stop in the middle of a book and pick up listening on another Alexa device. Because all Alexa-powered devices can sync over the internet, then you will not need to skip through chapters to get to the part you left off at. You won't even need to remember where you stopped. You will be able to simply begin listening on the Tap or the Echo or whatever other devices you may have in your home.

You can use a lot of commands to listen to audiobooks, but Alexa does not understand commands to bookmark or make notes, control the speed of her narration, and subscribe to other magazine or audio magazines.

Here are some commands that you can use to control Alexa's reading of an audio book.

"Read (name of the book.)"

"Play the book, (title)."

"Pause."

"Resume by book." This command can be used even when you have not listened to an audiobook recently. The most recent audiobook that you have listened to, be it a day ago or a month ago, will begin playing right where you left off.

"Go back."

"Go forward." The above two commands will allow you to jump thirty seconds forward or backward.

"Next chapter."

"Previous chapter." These past two commands will allow you to navigate through the audiobook by chapter name, as you would navigate through scene selection in a movie.

"Go to chapter (number)."

"Restart." Using this command will restart the chapter, not the whole book.

"Stop reading the book in eight minutes." You can change minutes to hours, and you can change the number to the time you would like. This can be useful if you like to listen to a story as you are falling asleep. However, you will also be able to quickly go back and find where you left off if you have a "sleep timer" on your audiobook.

Alexa is also capable of reading Kindle books. However, not all books have an audio component. You can see which of your Kindle books can be read by following these steps. In the Alexa App navigation panel, select **Music and Books.** Select **Kindle Books** then **Books Alexa Can Read.** The Kindle book can be read using text-to-speech technology, which is not always 100% accurate. However, this can be a great option if you are trying to clean, and you want something to which you can listen.

You can use all of the above commands with Kindle books with the following exceptions. You cannot set a timer for Alexa to stop reading the book in a certain amount of time. Also, when you are beginning the reading process, you may need to specify that you want to listen to a Kindle book. You can say "Read my Kindle book" or "Play the Kindle book, (title)."

Conclusion

Go ahead and enjoy streaming music and audiobooks on your Amazon Tap. If you have trouble accessing these features, look ahead to the Troubleshooting Guide chapter, and you may find some answers for your problems.

AMAZON TAP: BASIC SETTINGS AND SKILLS

To fully understand your Amazon Tap, you need to know the basics. You cannot begin controlling difficult devices and changing the skills if you have not even learned the basics. If you have not taken the time to thoroughly explore your Alexa App, now is the time to do so. This chapter will walk you through your Alexa App and a few basic commands that you can use to complete it.

When you open your Alexa App, you will see your home screen. You will see a few boxes or cards that will show you some of your last interactions or orders. This provides you an excellent way to review purchases that you made orally. If you accidentally made a purchase, you should see it appear in this menu, and you should be able to cancel it immediately.

You will see a little "hamburger" sign in the left corner of your Alexa App. This is your menu and will give you the options that you need to know in order to make many of the changes or adaptations that may be necessary to fit the Tap to your needs.

You should see eleven options in the drop-down menu when you click on it. These options are in the following order, and they include a brief description of their purpose and what you will see when you select them.

- Home- This option will take you right back to where you were before where you will be able to review your newest interactions with your Tap and other devices connected to the Alexa App. It is also a great way to monitor your children's activity if they have access to Alexa devices.
- Now Playing- This will show your music history. If your Tap is currently playing a song, you will be able to view the song's title and information. You will also be able to view upcoming tracks (if you selected an album or playlist) and the songs that have just played.
- Music and Books- This selection contains many different options for settings regarding playing music and access to music. You will use this tab if you want to change your default music library or streaming system. You will also use this tab if you want to find more music or audiobooks for your enjoyment on your Tap.
- Shopping and To-Do Lists- Making lists is one of the neat things the Tap can do via Alexa. You Alexa App comes with two default lists: one for shopping and one for tasks that you must complete. You can use some of

these commands to make Alexa add things to your lists. "Add (item) to shopping list," "Add (task0 to to-do list," or "Read me shopping list." You also have the ability to access the lists you have made from your tablet or computer, meaning that you can print them if you need to take your shopping list with you.

- Timers and Alarms- You will learn in the chapter entitled "Using Amazon Tap in Your Daily Life" how you can command the timers and alarms on your Tap to make yourself more productive. However, when you begin looking for where you will find those controls in the Alexa App, you will need to select this tab. There are a number of commands that you can make with your voice, but you will find that you do require the Alexa App to complete some functions that the voice-control cannot.

- Skills- In the next chapter, you will learn about skills that you can add to your Alexa App which you will be able to access from your Tap. These skill sets can expand your Tap's knowledge and ability. To access the skill packets that your Tap has, you can use this tab.

- Smart Home- This tab will be used to manage the smart home devices that you can connect to your Tap. In chapter three- Voice Control for Devices- you learned about how you can connect and control a smart home thermostat. You will access those controls from this tab.

- Things to Try- Because your Tap device is powered by Alexa, you will be able to use "Easter eggs." These are fun phrases that you can use to evoke a fun response from your Tap. There are lists of hundreds of them online, and you may find a few more in the Things to Try section. This section will also have some basic instructions on how to complete different tasks using your Tap. This tab will provide a great refresher for you when you are trying to remember how to listen to audiobooks, for example.

- Settings- This tab is one of the tabs that you will use the most. You will use this tab for setting up new devices or features often. Anytime you want to connect something else to Alexa to control with your voice, you will access the controls from here. You will also use this tab to access Alexa's voice training skills which you will learn more about in a later chapter. Last of all, you can use this tab to control basic things about your Tap such as the volume or location.

- Help and Feedback- This is the tab where you will be able to leave feedback about your Tap device. Also, if you have experienced problems with some area of your

Tap not functioning, you can talk to a representative who will answer your questions. You can also look through the guide that will help you work through any questions you might have.

- Sign Out- This is simple. If you are signed into another family member or a friend's Alexa App, you can sign out there so that you can access the Alexa App account that pertains to your Tap.

Those are the basic menu settings that you will find for your Alexa App that can be used to control your Tap. If you want to know about Alexa's more advanced settings and skills, continue on to the next chapter.

ADVANCED SETTINGS AND SKILLS

In this chapter, we are going to talk about some advanced settings and how to operate them. You will also learn how to add skills to your Alexa device and what skills are the best to have handy.

How To Add and Access Skills

Go to your navigation menu in your Alexa App. Select the **Skills** tab that was mentioned before. You will then be taken to the Alexa Skill Store. If you have the name of a specific skill set in mind, you can type it into the search bar. Or if you are simply looking for something new to add to your Tap's skills, you can browse through the categories and find a skill set that appeals to you. Once you have found the skill set that you want to use, you need to select **Enable Skill.**

Once the skill is enabled, you will be able to access it directly from your Tap. You can use some of the following commands to access skill sets: "Open (name of the skill)," "Disable (name of the skill) skill," or "(name of the skill) help."

The Best Skill Sets to Add To Your Tap

There are many, many skill sets to choose from, and new skill sets are published weekly. Here are a few skill sets that cross a variety of different tastes and might just be something that will change the way you view your Tap.

- The Seven Minute Workout- This skill set gives you a way to track your exercise at home if you don't have time to go to the gym. It is said that regular exercise every day is better than long bursts of exercise for one day that you do not follow up on. This skill allows you to choose from a variety of workout routines. You can then access your finished routines to track your progress and continue working on a workout plan that you set. The routines are set up to burn calories, not just stretch your muscles. This skill set is so useful that it is one of the most popular skill sets within the Alexa-enabled skills. You will mainly see this skill useful when you only have time for a short, calorie-burning workout.
- The Bartender- This skill set is one that will provide your personal drink recipes. Basically, this app has a variety of drink mixes that you may have had before and the recipes for those drink mixes. One benefit of this skill set is that the recipes are very accurate. You will be pleased with the result. Also, the variety of options provide a selection for everyone. You can use Alexa to

ask about a certain drink recipe, and she will be able to read you the recipe from this skill set.

- SmartThings- This is an excellent skill set if your devices need a smart home hub in order to operate properly. Instead of needing to purchase an expensive smart home hub, this skill set will bring the technology of a smart home hub with its own smart home hub. You can use this skill set to connect thermostats, light bulbs, sockets, and more. Users of this product have said that the set-up is quick and easy and that it is produced by a trusted brand, meaning that you will have guaranteed satisfaction in the skill's performance.

- Akinator- This skill set is purely for fun. It is the game of twenty questions but with Alexa. You can sit there with your Tap and play twenty questions. You will be amazed at how accurate Alexa can be when she is guessing who you are thinking of.

- Capital One- Capital One is a skill set that is a bit different from the rest. It is not just created for fun. It can also be a quick and useful way to check your bank account. You can look up your credit card balance and pay your bill. You can also check on your loans, all through Alexa. This app will make it harder to forget to pay your credit card bill. You can simply access what you owe by asking Alexa a question.

- Automatic- This skill set only works if you have already purchased the Automatic device. However, if you have the device, then you will have access to multiple neat features. You will be able to ask about your car's location and gas level, which can prepare you if you need to leave early in the morning to buy gas. You can also ask about the number of miles that have been driven in the last week or month or even on a specific day.

- Domino's Pizza- Last of all, this skill set will make ordering your pizza easy. You will need to have your domino's account and credit card information stored within your Amazon account to complete this. However, once the skill set has been installed, you can order the pizza that you want in a much easier way than calling in. You can even ask Alexa about the status of your pizza to see when it will be delivered.

Alexa's Advanced Settings

If you navigate to the **Settings** tab on your Alexa App menu, then you will be able to access these advanced settings features. There are multiple options that you can access and control, but we will only be discussing the most important at this point. Here are the ways that you can access these settings and adjust them.

- Location- Once in your settings tab, you should find a section that says **Device Location.** Select that tab, and you will be able to enter your device's location. You can normally enter your home address. You can use this address to check your commute; Alexa will be able to tell you if the route has a lot of traffic or not. You will also be able to check the weather pertinent to your location. Another way you can use your location is to look at movie show times or activities in your area.
- Battery- Once you are in your settings tab, you will see a section entitled **Battery.** This will allow you to check your device's battery level. You can also ask your Tap "What is your battery level?"
- Calendar- You may choose to link your Google calendar to your Tap. Google calendar is currently the only calendar system that is supported by Alexa. By going into the **Calendar** section which can be found in your settings, then you will be able to choose **Link Google Calendar account.** You will need to sign in with your Google account information. If you don't have a Google account, but you would like to make a calendar, you can

create a free account. Any events that are on your calendar will automatically become accessible to Alexa. You can use the following phrases to complete actions with your calendar: "when is my next event?", "what's on my calendar?", "What's on my calendar tomorrow?", "Add an event to my calendar," or "Add (activity) to my calendar for (day) at (time."

Those are some of the advanced settings that will affect the usefulness of Alexa. If you don't put in your location, for example, Alexa will not know the weather in your area. Putting the correct information in these areas can help your Tap be able to give you useful information.

USING AMAZON TAP IN YOUR DAILY LIFE

There are many ways that your Amazon Tap can help change your daily life. You have already learned about how you can add and manage a Google calendar from your Tap. You have also learned about how you can add skill sets to Alexa. All of those will affect your daily life; however, there are a few more areas that should be addressed as well. In this chapter, you will learn how to manage alarms and timers, how to shop with Alexa, how to hear about the news and traffic, and how to order and check on an Uber with your Tap.

Manage Alarms and Timers

Managing alarms and timers using your Tap is not difficult. Most of the functions within alarms and timers can be completely controlled by your voice. However, there are a few things, such as canceling repeating timers that you will need the Alexa App to do. Here are some basic commands that you can use to manage alarms and timers.

"Set an alarm for (time)."

"Set an alarm for (amount of time) from now."

"Set a repeating alarm for (day of the week) at (time)."

"Set an everyday alarm for (time)."

"Snooze." This is a nine-minute snooze.

"What time is my alarm set for?"

"What alarms do I have for (day)?"

"Stop the alarm." This command will stop the alarm from sounding when it is going off.

If you want to delete an alarm, you will need to use the Alexa App. You will go into the **Timers and Alarms** section that was discussed earlier. You will see a list of your Alexa-enabled devices. You will select your Tap, and you will choose the **Alarms** button. Here, you will be able to view all of the alarms that are set to go off for this device. Choose the alarm that you want to delete and press **Delete Alarm.**

You can also change the repetition of the alarm to a more specific time frame using the Alexa App. You will follow the same steps above to enter and view the alarms that are set. You will select an alarm and click **Repeats.** You can choose to never have it repeat or to have it repeat on the weekends or the weekdays or every day.

If you want to check on the status of a timer, you can say: "How much time is left on my timer?" You can have multiple timers set, and your Tap will tell you how much time is left on your upcoming timer.

Shop with Alexa

Your Tap will be able to help you place orders for items or music, and you will even be able to track these items just using your voice. Placing orders is easier for your Tap to complete if you have already placed the same order before or if the order is digital. For example, if you are just looking to order a pair of jeans, you would not want to do that through voice shopping as you would have no idea if they are baggy or tight. When you shop with Alexa, she can inform you of the product name, price, and approximate arrival date. She cannot read you the description or the reviews, meaning that voice orders should stick to things will which you are comfortable ordering. If you have a Prime membership, Alexa will look through Prime eligible items first, so that you do not need to pay shipping if you do not have to do so. You can also use Alexa to add an item to your cart in case you want to view the item before you make the purchase.

You can manage your ability and your family's ability to make purchases using your Tap by controlling the options on the Alexa App. Under the settings tab, you should see something that says **Voice Purchasing.** You will click on that and be able to change the following options. You can select to turn on or off voice purchasing. You can enter a four-digit confirmation code that must be entered before a purchase can be made. This will help you confirm that you want to make a purchase before accidentally ordering something. Also, you will be able to change your 1-click billing information.

You can also shop electronically and purchase music using Alexa. You need to make sure that you have a billing address in the United States and bank within the United States from which Amazon can take the money. If you have this, then you can virtually order music from any part of the world. You can use the following phrases to shop for music.

"Shop for the song (name of the song)."

"Shop for the album (name of the album)."

"Shop for songs by (name of the artist)."

"Shop for new songs by (name of the artist)."

"Buy this (name of the song or album)."

"Add this (name of the song or album) to my library."

If you already have a Prime account, some songs will be available at no extra cost to you. Amazon will ask you if you want to make the purchase, alerting you that the song is already in the free Prime library. This will keep you from making an unnecessary purchase. You can also say "Track my order" to learn the location of your nearest order.

News and Traffic

To receive your news every morning, you can choose from a number of popular news stations. Alexa will read off the latest pre-recorded news update to keep you in the know of the most recent happenings. You can choose the news network that gives you the update from this list: NPR, BBC News, the Economist, and the Associated Press, along with weather from AccuWeather. In order to receive your news in this way, you will need to set up this function.

In your settings menu, you should have an option that says **Flashbriefing.** You can select how many networks that you want to provide you with news updates. You will mark each area as "on" if you want them to be a part of your flash briefing. You can also choose **Edit Order** to choose in which order you want the networks to play their information. For example, you may want the weather to be played first.

These are the phrases you can use to control your flash briefing.

"What's my flash briefing?"

"What's new?"

"Next."

"Previous."

"Cancel."

The last three orders were designed to navigate within your flash briefing. Because you can set up the briefing to give you news from different networks, using the "next" or "previous" commands will help your Tap skip between news networks. The "cancel" command will stop your flash briefing altogether.

Make sure that your address is accurately set to find out the traffic in your area. If you do not know how to set your address, look in the previous chapter. Once your address is set, you can ask any of the following questions to get an update on the traffic in your area.

"How is traffic?"

"What's traffic like right now?"

"How is my commute?" If you put in the address of your job, you will be able to ask about the route to work. Your Tap will be able to tell you if it is unusually long or very fast. This can help you prepare for when you need to leave in the morning.

Order an Uber Driver

Uber has recently become a popular smartphone app to order a ride or food delivery. Uber has your bank account information, and there is no need to carry cash or hand cash over to a driver who may charge more than you thought. However, by connecting Uber to your Amazon Tap, you can actually order a car just by using your voice. That simplifies the process even more.

Uber will be located under the **Skills** tab on the menu of your Alexa App. Search for Uber and then choose to "enable skill." You will need to create an Uber account if you do not have one or sign into your account if you already have one. Once the skill has been enabled, you can control it with your voice. You can use these phrases to control your Uber transportation.

"Ask Uber for a ride."

"Ask Uber to order an Uberblack."

"What is the status of my Uber ride?"

"Cancel my Uber ride."

With these key phrases, you can order a ride, cancel a ride, and check the status of your ride, changing the way you get ready.

TROUBLESHOOTING GUIDE

Here are a few common problems that users have with the Amazon Tap. If you do not find a solution to your problem here, you can always look in the **Help and Feedback** section of your Alexa App menu.

Trouble Connecting to Wi-fi

Here are a few things you can try to fix your wi-fi connection if it does not connect. Firstly, try again following the same steps that are outlined in chapter two. Next, see if the problem is related to your wi-fi network rather than the Amazon Tap by trying to see if other devices are having trouble connecting to the network as well. Your wi-fi could be experiencing a temporary problem, and you can try to connect again later. Last of all, try updating the firmware for your modem. It may be outdated and need the latest updates to connect correctly.

Your router's placement and use could be affecting your ability to connect. Try moving your wi-fi router out of an enclosed space and away from objects that emit radiation, which will interrupt the signal. Try moving your Amazon Tap closer to the wi-fi router, and turn off any devices that you aren't using to see if there is simply too much congestion on the wi-fi network.

Amazon Tap Charging Problem

If you are having trouble charging your Tap, the problem may be that you are using a power adaptor that is not meant to be used with the Tap. Check that you are using the power adaptor that came in the box. If you are trying to charge the Tap from the USB port on your computer, it will take a lot longer, making it seem as though it is not charging properly. Always plug the charger into the wall.

The Power button should glow when charging; however, if the Tap was completely dead, you might not see the button glow for a few minutes. Try taking out and reattaching each part of the charger. Wiggle the parts gently to see if one part is especially loose. Last of all, you can try restarting your device. If none of these methods work, try calling for assistance.

Connection to Smart Home Device Not Working

If you are having trouble connecting your Tap to a smart home device, look online to make sure that the smart home device you have purchased is compatible with Alexa-enabled devices. Your smart home device should have a companion app or skill set that you can download. Downloading the skill set should make access easier. Try restarting both your Tap and the smart home device. Sometimes, restarting the device can help your Alexa App "discover" it and be able to connect.

Make sure that you have fully updated your Tap and your smart home device. Without the updates, the two devices might not be able to communicate.

Bluetooth Connection Problems

Make sure that your Bluetooth device has one of these profiles: Advanced Audio Distribution Profile (A2DP SNK) and Audio/ Video Remote Control Profile (AVRCP). Without these two profiles, your device may have trouble streaming audio from your Tap or mobile device, and you may have trouble using hands-free control.

Make sure that your Bluetooth device has been fully charged. If your device has a non-removable battery, then make sure it has been charged. If your device runs on replaceable batteries, put in some new batteries to see if the connection problem was simply related to your Bluetooth device not having the energy to talk to your Tap.

Another option is for you to try pairing the devices again. When you pair them, make sure that the devices are fairly close. A lot of space can make it difficult for the devices to find each other. Also, find an area that does not have a lot of other wireless devices as the signals may become confused. If you have too many Bluetooth devices connected, you may have trouble connecting another. Help your Tap "forget" these

Bluetooth devices so that it can focus on the Bluetooth device with which you want to connect.

Go into the setting tab of your menu in the Alexa App. Choose your device, the Amazon Tap, then select **Bluetooth.** You should see a button on the screen that says **Clear.** Press that button. Now, you can pair the devices again. Make sure that your Bluetooth device is on, then tell Alexa "Pair." This should connect the devices.

Conclusion

Remember that if you have more problems that were not covered in this troubleshooting guide you should look in your Alexa App menu. The Alexa App was created to complement the hands-free devices and will have plenty of help sections for any problem you might have.

VOICE TRAINING

In the beginning, your Amazon Tap may have trouble understanding your voice. She was created to learn over time to adjust to your voice. However, if you want to speed up that process, you may choose to participate in voice training with Alexa. In this chapter, you will learn the basic steps of how you can voice train Alexa in your Tap.

Prepare for Your Training

Before you begin the voice training session, you want to make sure that you are going to use the voice training as you would if you were normally talking to Alexa. If you normally talk to your Tap at a rapid speed or slurring your words together a little bit, then do that when you are voice training. Don't try to especially annunciate your words unless you normally speak like that. Also, turn off any background noise that will confuse your Tap.

Start Voice Training

Open your Alexa App and select **Settings** from the drop-down menu. You should see a tab entitled **Voice Training.** You should be able to select your Tap from a list of Alexa devices that are connected to your account. Once you have chosen a device, click **Done.** Honestly, if you have more than one device, it does not matter if you train Alexa on a different device. Your voice patterns will be saved onto all the devices that are connected to your account. However, training your Tap specifically can help you learn how you need to speak to your Tap to help it understand you completely.

Sentences will appear on your Alexa App, and your Tap device will be ready to listen. You need to press the microphone button and read the sentences off your device in your normal voice. Remember not to slow your voice unusually. Once you have read a sentence, you can select **Next** to be given the next sentence. If you read a sentence and it completely comes out of your mouth wrong, you can repeat the sentence. Select **Pause** then **Repeat Phrase.**

Here is a tip about working through the voice training session with your tap. Read the sentence off the screen silently then look away and repeat the sentence. Many people have two different voices: one is used when they are reading aloud while the other is used when they are saying commands in normal life. In order to really train your Tap to respond in the best way possible to your voice, you will need to practice the command as naturally as possible. You do not want the Tap to understand your reading aloud voice only and not your everyday speaking voice.

Each voice training session will provide you with twenty-five statements. After you have spoken the last one, select **Complete.** You will then be given two options. You can **Start New Session** or **Go To Home Page.** If you want to work on training your Tap even more, you can choose to go through another voice training session. However, if not, you can simply close out your session.

Does Alexa need more than one voice training session? In reality, probably not. Unless you were using your reading aloud voice, she probably understands you pretty well after one session. It is good to note that the prompts for each voice training session are different. The voice training sessions might help you discover new ways that you can command your Tap, making the voice training sessions educational for you.

CUSTOMER TESTIMONIAL– SARAH LATSHAW

"It seems that every few months, a new 'must-have' tech gadget arrives on the market. When Amazon announced the release of their new Tap, a portable Bluetooth speaker with integrated Alexa Voice Service (and a list price of $129.99), to say I was skeptical would be an understatement. Portable Bluetooth speakers are hardly a new technology, and many of them are far less expensive. My skepticism only grew when the speaker arrived. Something so small and understated could hardly stand up to the bold claim of 'crisp sound powered by Dolby.' Then I turned it on… and my mind was changed in an instant.

The quality of sound produced by the Amazon Tap is nothing short of amazing. The deep resonance of its bass is perfectly accentuated by treble notes that are barely perceptible on most other speakers. Listening to music on this device is something that can best be described as a highly immersive experience. It can play notes that you may have forgotten were included in songs you hear on a regular basis. Spoken words are smooth and well-balanced with great definition. The superior bass and treble are not accompanied by any of the cracks, pops, or buzzing that plague many similar devices. Its omnidirectional sound, powered by dual speakers, makes it a perfect choice for entertaining a crowd.

In terms of grab-and-go convenience, there are obviously smaller and more portable speakers on the market; but the Amazon Tap is far from clunky. It is shorter than a standard pencil, with a diameter the same as a soda can. This makes it the perfect size to fit in the drink holder of your car, backpack, diaper bag, or even a medium-sized purse. Charging it is very simple, and the battery will support continuous playback for up to nine hours. The included charging cradle has a very low profile, adding just over half an inch to the device's height and nothing at all to its diameter. The cradle has a micro-USB input and its two points of contact ensure that the device will always connect to its charger no matter which way it is facing. If you don't have the charging cradle with you, the Amazon Tap can still be charged by connecting a micro-USB cable to the port below its power button.

With all of these stellar features, there are still a few drawbacks. You have to push a button in order to interact with its Alexa interface, which can be frustrating if you are looking for a totally hands-free environment. Though it is very compact, it is still not the smallest speaker available, so you can't just put it in your pocket. It is water-resistant but not waterproof, so you have to remember to be careful when you're out and about with your device. The most frustrating thing about it is that you can't interrupt it when it's talking; if it wants to tell you for the umpteenth time how to connect your Bluetooth, or give you a sales pitch for Amazon Music, you have to listen. If it responded to "stop, Alexa" or something similar, that would be ideal. Not exactly a deal breaker though.

I expected to find the Amazon Tap to be a frivolous product, but I was quite pleasantly surprised. This tiny device produces rich, full-bodied sound in a highly portable format. If you purchase a Sling case, it becomes even easier to transport. Affordable, high quality sound on the go? Amazon Tap has it all.

CONCLUSION

The Amazon Tap is an amazing development from the Amazon company. It provides a lot of different features that the Amazon Echo and the Amazon Echo Dot do not have. Here are some of the best features about the Amazon Tap.

First of all, the Amazon Tap is operated by a button. There are a few reasons for this button. It means that the battery is not drained as quickly. Because the Amazon Tap is a battery-operated device, this means that if the Tap was constantly listening, the battery would be drained very quickly. However, unless you are using the Tap to stream music or speak, it is not losing battery. Also, any privacy concerns about having the Echo and the Echo Dot always listening have been addressed by the Tap that only listens when you want it too.

Next, the Amazon Tap is battery operated, as stated earlier. This means that the Tap truly is a great portable speaker. You can connect the Amazon Tap to mobile hotspots or wi-fi and receive great sound from the speaker no matter where you go, only needing to charge the battery every nine hours.

Last of all, the Amazon Tap's price falls at a great middle ground. It can do everything the Amazon Echo can do, but it can do them at a much lower price. While the Amazon Tap is more expensive than the Amazon Echo Dot, the Echo Dot also does not come with a quality speaker. With the Amazon Tap, you can have the quality speaker and the Alexa benefits all in one place.

Remember that you can always answer any questions you might have with the help section in the Alexa App. I hope that you now feel fully prepared to operate your Amazon Tap and enjoy it to the fullest!

www.ingramcontent.com/pod-product-compliance
Lightning Source LLC
La Vergne TN
LVHW050148060326
832904LV00003B/52